Book Title:

From YouTube to Fake Tube: The Freedom to Disable Artificially Generated Content on YouTube and other Social Media Applications

Creator of the 2013 Goal Setting Workbook,

'The Cycle Case Number Goal Invention System'

Author of the 2020 book,

Goal Invention: The Mental Therapy, Sport, and Hobby

Disclaimer: This is an opinionative book. The opinions expressed in this book are entirely my own.

Table of Contents

Chapter 1: Introduction

I wrote this book to raise awareness about a very frustrating and negative reality that is occurring on YouTube and other social media applications.

I wrote this book to explain the real issue of artificially generated fake videos flooding and spamming the YouTube algorithm on the home page sections and also spamming the videos that appear on the related videos' sections of YouTube as well. The spamming of artificially created videos affects not only YouTube, but all social media applications.

Artificial intelligence-generated videos are created from programs. Often, a multitude of programs are used to create a fake video.

The most common form of artificially synthetic videos is where a program generates a script. The script is then read with a life-like realistic robot voice.
The then fake script and voice narration is added to either pictures or videos.

Chapter 1: Introduction Continued.

A fake video often uses real life footage or pictures rather than artificial synthetic images/videos, as it makes it less obvious to tell that it is a fake video thus increasing how many views the fake video gains. Programs create synthetic scripts and voices that are generated for synthetic videos, but the videography in the videos are of real-life footage and real life pictures.

In some cases, videos are entirely artificially fake synthetic videos, where the entirety of the videos are generated by a program where all the scripts, narrations, photos, and videos are entirely fake and generated by a program. Artificial fake videos spam, flood, and essentially bury YouTube and other social media applications into digital junkyards of synthetic fake content that are full of rubbish and pointless synthetic recycled videos created by programs.

The reality of YouTube and other social media applications being a dumpster fire of fake artificial synthetic videos finally struck me when friends of mine where fooled with fake videos that were about celebrities that used the previously mentioned method of synthetic scripts and synthetic voices played in the background of real life photos and real life videos.

Chapter 1: Introduction Continued.

The most common type of fake artificially synthetic content as previously stated is the usage of robot scripts and robot voices created by programs. I told my friends that the videos that they were watching were made by programs, and I then advised my friends to look at their home pages of recommended videos.

I showed my friends the recommended videos on their home page sections as a way to show just how much of the videos recommended are actually fake and artificially generated. Even while I was showing my friends how polluted the YouTube homepage & recommended videos algorithms were, I was struck back at how bad the situation had gotten.

To my shock, the majority (maybe more than 75%) of videos recommended on my friends' home pages and recommended videos sections were videos that were artificially generated synthetic fake videos.
That was the moment when I decided to call YouTube "Fake Tube". That was also the moment when I knew that I was needed to write a small book raising awareness and providing solutions to address this issue.

Chapter 1: Introduction Continued.

When the majority of videos on YouTube are not created by content creators, but rather generated by an artificial intelligence-program, it is no Longer YouTube and instead is “Fake Tube” or “Robot Tube”, respectfully.

What was once a platform for you, as a content creator, is now a platform for artificial intelligence - programs to spam, junk, pollute, and make it nearly impossible for genuine content creators to be found in the algorithm of YouTube.

“Fake Tube”/”Robot Tube”, oh sorry, I mean YouTube, was once an acceptable platform for content creators prior to the prevalence of click bait themed - artificial fake videos spamming the application’s home page sections and recommended videos sections.

In 2025, “Fake Tube”/ YouTube has become an abandoned junkyard wasteland of synthetic videos created by programs.

Chapter 1: Introduction Continued.

YouTube, in my opinion, is an application where content creators and users are treated with disrespect, disregard, and negligence by YouTube, the Company.

The act of YouTube, the company, promoting the spamming of the home pages and recommended videos sections with fake videos, can be described as treating the users of YouTube and content creators with disrespect and disregard for both content creators and regular watchers of YouTube.

The undeniable favorability that YouTube gives to fake videos swarming the algorithm of home pages and recommended video sections is a clear example of how YouTube, the company, does not care about the experience of content creators on YouTube or the users of YouTube.

With every low quality, painful to watch fake generated video that takes a spot on the home pages and recommended videos sections' algorithms, an opportunity for a genuine human content creator and quality watcher experience is taken away.

Chapter 1: Introduction Continued.

The immoral reality of "Fake Tube"/"Robot Tube"/ YouTube is not only an act of negligence to humans as content creators, but it is also an act of disrespect for users of YouTube.

As humans in a democracy, we are entitled to watch and browse for videos that are real videos of quality made by people and not generated by robots/programs.

Solution: There is a very easy solution to resolve this issue. In this small mini book, I will instruct my solution not through censorship but through the beautiful word of "settings".

The word "settings" will be the most beautiful word on YouTube and other social media applications in the future. In this book, I do not advocate for the censorship of fake videos on YouTube, as that is an area of topics where law makers must navigate the legalities of synthetic content and is beyond my expertise range.

In this book, I simply advocate for freedom of media consumption and settings configurations.

Chapter 1: Introduction Continued.

My second textbook which I wrote in 2021, called *"How Cellphones Cropped, Distorted, and Visually Censored Cinema: Let's Save Cinema"*, also advocates for the freedom of media consumption of consumers to watch movies and videos on YouTube, Netflix, Hulu, Prime Video, Tubi, in their original full screen 4:3-16:9 aspect ratio versions in addition to their ultrawide 19.5:9 - 21:9 aspect ratio versions.

Freedom of media consumption is a big deal to me, and the issues discussed in this mini book also pertain to the topic of freedom of media consumption.
My solution to resolve the issue of YouTube being a junkyard of fake videos is through creating a new setting feature, that would be accessible on YouTube and other social media applications both while being signed in and while signed out of each respective application.

I propose this settings solution to be termed the "FDAIG" setting, the "Freedom to Disable Artificially Generated Content" setting button or the "Freedom to Disable Artificial – Intelligence Generated Content". I use both wordings throughout this book.

Chapter 1: Introduction Continued.

This setting would be accessed and seen with a large icon button titled “FDAIG”. The “FDAIG” icon setting button would be easily visible and accessible in the upper left or right corners of YouTube and other social media applications.

Note: In this book I use the term “act” in some parts, as I believe that this setting should become an FCC compliant law in the future.

The remainder of my book will include reasons and answers

End of Chapter 1.

Chapter 2

Table of Contents of my reasons:

Chapter 2: Reasons to Apply the FDAIG Setting on YouTube

Why the "Freedom to Disable Artificially Generated Content" or "FDAIG" setting icon act is important to apply on YouTube and other social media platforms.

Reason 1 : Because the YouTube Creator and Consumer (the Watcher) Experience is Rubbish

Before 2023, YouTube was a predominantly effective tool, experience, and platform for content creators.

From 2006 – 2022, humans could create videos to promote their brand and crafts on YouTube with a relative efficiency and ease of attaining an audience.

Beginning in late 2023, and particularly 2024, I noticed an alarming trend. A trend of pollution and littering of YouTube with fake generated videos.

The trend of artificially generated content destroyed the usability of YouTube where now the home pages' algorithms and recommended videos' algorithms are no longer functioning or even usable for consumers.

Chapter 2: Reasons to Apply the FDAIG Setting on YouTube

Reason 1 Continued.:

At the time of writing this mini book in 2025, my patience with YouTube the company is gone, as the user experience of YouTube is absolutely spammed to the brim with rubbish fake videos.

When users can no longer use the home pages and recommended video sections of YouTube, the experience of discovering new videos that are not fake is greatly reduced and the experience is horrible as a result.

The YouTube experience from the consumer/watcher standpoint, is about as enjoyable as a strong headache without any medicine. The click bait fake videos could be used as a form of legalized torture.

Chapter 2: Reasons to Apply the FDAIG Setting on YouTube

Why the "Freedom to Disable Artificially Generated Content" or "FDAIG" setting icon act is important to apply on YouTube and other social media platforms.

Reason 2 : Because the YouTube Content Creator Experience is Poor, Non-Incentive, and Unfair.

Another major issue of the dumping of fake synthetic videos on YouTube's algorithm is that it makes it very difficult for human content creators to earn a full time living or part time-living, let alone gain an audience of viewers of any kind.

For every fake video that takes a spot in the algorithm, the spot of a real human content creator is taken away.

In 2025, the majority of videos on YouTube's algorithm are trashed with fake videos spewing negative, random, garbage, purposeless, click-bait, biased, recycled, and/or propagandic related contents.

Chapter 2: Reasons to Apply the FDAIG Setting on YouTube

Reason 2 Continued.:

Many of these fake videos on YouTube are blatantly misinformative, click-bait, biased, propagandic, or hateful videos that target celebrities, politicians, or any high profile figures that will gain the most viewership and ad revenue.

In 2025, humans who are content creators do not have a chance for their content to be enjoyed by humans. YouTube should be for you, not for some anonymous accounts from parts unknown and locations unknown – that release trashy fake videos.

These fake videos could literally be considered a form of mental torture, as they trash the algorithms of YouTube users' home pages and recommended videos sections.

The complete and utter disregard by YouTube for content creators and content consumers shows that the company is no longer YouTube, as it is not a platform for You.

YouTube is not a platform for you, the content creator. YouTube is not a platform for you, the consumer.

Chapter 2: Reasons to Apply the FDAIG Setting on YouTube

Reason 2 Continued. :

YouTube is a platform that prioritizes fake videos from anonymous accounts from anonymous locations. YouTube does not care about you whether you are a content creator or an average YouTube user/watcher.

On the bright side, from 2006-2022, YouTube lived up to its name as it provided a good way for content creators and consumers to broadcast their own content on their channels and grow their viewership effectively over time.

As of 2024 and especially 2025, YouTube is now “Fake Tube” or “Robot Tube”: A digital landscape where artificial intelligence-programs go to dump waste in the form of synthetic content with fake scripts, fake pictures, fake robot voices, and/or fake videos.

“YouTube is not a platform for you, the content creator. YouTube is not a platform for you, the consumer.”

Chapter 2: Reasons to Apply the FDAIG Setting on YouTube

Why the "Freedom to Disable Artificially Generated Content" or "FDAIG" setting icon act is important to apply on YouTube and other social media platforms.

Reason 3 : To Save YouTube and other Social Media applications in general.

YouTube (and other social media applications also) is now an application in shambles and is contaminated with an algorithm that is a nonfunctioning fake synthetic polluted junkyard. The user experience is enough to make a grown person want to destroy whatever device the individual is watching YouTube on.

The reality of YouTube as of 2024, is the death of a platform, a platform that diminishes creativity by creators for consumers, by spamming the platform with fake videos generated by programs. The spamming of fake artificially generated videos on YouTube destroys the content creator and consumer experience.

Chapter 2: Reasons to Apply the FDAIG Setting on YouTube

Why the "Freedom to Disable Artificially Generated Content" or "FDAIG" setting icon act is important to apply on YouTube and other social media platforms.

Reason 4 : To Encourage Authentic Quality Videos Made by Human Content Creators rather than Trashy Fake Videos Generated by Programs

A man once said, "there is nothing more moving, powerful, and purposeful than a beating heart" - me. The current algorithm greatly diminishes the livelihood of human content creators, which removes the incentive for quality video productions.

From 2006 – 2022, the YouTube home page and recommended videos sections were effective.
From 2006 – 2022 the YouTube algorithms promoted videos that were created by human content creators for human consumers. Any one lucky enough to be alive from 2006 - 2022 will remember all the amazing content creators that shined and economically thrived on YouTube.

Chapter 2: Reasons to Apply the FDAIG Setting on YouTube

Reason 4 Continued.:

In fact, the further you go back in time, the higher in quality, interest, and entertainment the platform of YouTube was.

YouTube was YouTube for a decade and a half but for the last few years it has turned into “Fake Tube” or “Robot Tube”, respectfully. A platform where every usage of the home page or recommended videos feature causes instant headaches and borderline mental nausea.

The prevalence of fake synthetic videos on YouTube’s home page sections and recommended video sections destroys content going forward, as quality human made content will never be seen or found because of the trashing of fake videos on YouTube’s algorithm.

The only feature that is still functional on YouTube since 2024 is the subscriptions feature.

Chapter 2: Reasons to Apply the FDAIG Setting on YouTube

Why the "Freedom to Disable Artificially Generated Content" or "FDAIG" setting icon act is important to apply on YouTube and other social media platforms.

Reason 5 : By Promoting Freedom of Media Consumption for Adults, and Providing Parental Choice for Parents of Kids

As a grown individual over the age of 30, I can manage the effects of fake videos or synthetic videos that spam the YouTube algorithm and other social media applications as well.

However, the effects of artificial intelligence-created videos with synthetic scripts, pictures, or the videos themselves are most likely horrific to the mental wellbeing and development of humans including adults and children. The majority of fake videos on YouTube are hateful, propagandic, biased, agenda-based, click-bait, recycled, and generally negative, unenjoyable, and pointless in nature. I would not wish fake videos on my worst enemies.

Chapter 2: Reasons to Apply the FDAIG Setting on YouTube

Reason 5 Continued.:

The majority of fake videos on YouTube are created with robot scripts and life-like robot voices videos that spam our timelines from anonymous accounts from anonymous locations from anonymous electronic parts.

The fact that there is not a clear setting icon button like the proposed "FDAIG" 'Freedom to Disable Artificial Intelligence - Generated Content' icon button, is a blatant violation of freedom of media consumption and a violation against the rights of parents to prevent their children from the harms of synthetic content.

Under basic parental laws, every parent and human should be provided with this proposed "FDAIG" icon setting button, the 'Freedom to Disable Artificial Intelligence - Generated Content" icon button that would display on YouTube and other social media applications in large font on the upper or lower corners of the screen.

This "FDAIG" setting icon button would be accessible while being signed in or without being signed in to YouTube or other social media applications.

Chapter 2: Reasons to Apply the FDAIG Setting on YouTube

Reason 5 Continued.:

If I, a grown man feel like I'm being brainwashed, violated, and tortured by synthetic videos spamming the home pages and recommended videos of YouTube, how will a child, teenager, or young adult manage this situation without their quality of life being negatively affected.

Many parents want their children to not be raised by robots from parts and locations unknown. To be against a proposal like the "FDAIG" setting icon button to disable fake content on applications is an inhumane viewpoint, as it rejects the right of parents to obtain a setting that disables fake content for themselves or for their children.

Parents should be legally able to control the settings of synthetic fake content that their children are able to consume on YouTube and other social media applications. Not providing the public and parents with a clear setting icon such as the "Freedom to Disable Artificially Generated Content" icon button shows lack of empathy or general concern for humans on the part of YouTube and other social media applications.

Chapter 2: Reasons to Apply the FDAIG Setting on YouTube

Reason 5 Continued.:

It is not an exaggeration to say that the experience on these platforms such as YouTube and other Social Media Applications are hazardous.

Humans have never been more miserable mentally (the invisible statistic) and forced artificially generated synthetic content is one of the main reasons of suffering. Another reason for human suffering is the excessive usage of YouTube and other social media applications.

Look at the vast majority of fake videos on YouTube, and ask yourself, 'would you wish you were raised on those type of videos in your childhood?'

Obviously, most of the answers would be no.

The prevalence of fake videos on YouTube and other social media applications, without a setting to disable them, shows lack of morality and commonsense from both secular and religious perspectives.

Chapter 2: Reasons to Apply the FDAIG Setting on YouTube

Why the “Freedom to Disable Artificially Generated Content” or “FDAIG” setting icon act is important to apply on YouTube and other social media platforms.

Reason 6: Improving Mental Health of Children, Teenagers, Young Adults, and Adults

How can we expect the humans of today and tomorrow to be physically and mentally happy and healthy when their brains are constantly engaging and interacting with artificial synthetic videos created by programs.

A wise person once said, “the mind is a terrible thing to waste”. The usage of our brains is our greatest gift and ability. As humans, it should be our fundamental right to choose to engage our brain with content created by humans (and not programs) on YouTube and other social media platforms.

Chapter 2: Reasons to Apply the FDAIG Setting on YouTube

Reason 6 Continued.:

This is our choice, and it is our right. This is freedom of media consumption.

The world is at a massive crisis point. I believe that humans, adults, and children alike, deserve the basic right of freedom of media consumption, the freedom to configure settings on YouTube and other social media applications in order to disable artificial synthetic created videos and content. This includes videos that use artificially generated scripts and robot voices (often fake videos will play real footage over fake scripts and fake voices).

The failure of YouTube and other social media applications to provide users with the ability to disable fake synthetic videos is inhumane and violates the basic right of freedom of media consumption to configure settings on applications.

Chapter 2: Reasons to Apply the FDAIG Setting on YouTube

Why the "Freedom to Disable Artificially Generated Content" or "FDAIG" setting icon act is important to apply on YouTube and other Social Media Platforms.

Reason 7 : to Promote Creativity and High Quality Content, Productions, and Entertainment on YouTube and other Social Media Platforms.

The spamming of synthetic videos on YouTube's homepage algorithm and recommended videos algorithm reduces creativity, reduces quality content, and reduces quality productions.

Here's Why:

Artificial Intelligence-generated content kills creativity, for the fake-trashy content takes up much of the space of the algorithms on YouTube. With less of a chance for human content creators' videos to appear on peoples' home pages and recommended videos on YouTube, there is less of a financial incentive for content creators to produce quality videos.

Chapter 2: Reasons to Apply the FDAIG Setting on YouTube

Reason 7 Continued.:

It is also virtually impossible for new content creators to establish audiences as their opportunities are greatly diminished due to the spamming of fake videos in the YouTube algorithms.

There is very little financial incentive, and very little motivation for new content creators to use YouTube and other social media platforms: creativity no longer has an incentive on YouTube and other social media applications (as many featured video spots on the algorithm are taken by fake generated videos).

Where artificial intelligence begins, creativity ends, and the incentive for human creativity ends.
Don't tell me that the 2025 YouTube experience is better than the YouTube experiences of 2006, 2007, 2008. The farther you go back in the timeline of YouTube from 2025 backwards, you will remember and realize that the content was more creative, unique, high-quality, and there were many amazing productions (from 2006 - 2016).
YouTube is a prime example of the statement **"Where artificial intelligence begins, creativity ends."**

Chapter 2: Reasons to Apply the FDAIG Setting on YouTube

Reason 7 Continued.:

The home pages and recommended videos sections of YouTube are now non-functional synthetic waste dumping NO-GO zones.

Synthetic content has made the YouTube homepages & recommended videos toxic no-go zones of program generated waste.

Installing the proposed "FDAIG" / "Freedom to Disable Artificial Intelligence - Generated Content" setting icon button would allow consumers and content creators to be in charge by allowing them the choice of disabling artificially generated content on YouTube and other social media platforms.

This simple settings configuration option would provide the incentive for human content creators to thrive and create enjoyable high quality entertainment. The "FDAIG" setting icon button would restore the incentive of people to create content for either financial reasons or simply for the purpose of building an audience.

Chapter 2: Reasons to Apply the FDAIG Setting on YouTube

Reason 7 Continued.:

Without an audience, there is little incentive for content creators to create high quality videos, and productions. The paragraphs from the prior few pages cannot be more true than simply looking back year by year at the content creator experience of YouTube and the consumer (watcher) experience of YouTube.

The first decade of YouTube (2006-2016) created legends filled with amazing people, amazing careers, and scores of breakthroughs and accomplishments that impacted and changed the world for the better. The YouTube of 2023-2025 does not provide the incentive, platform, or ability for legends to emerge.

The "FDAIG" / "Freedom to Disable Artificial Intelligence – Generated Content" setting icon button is a win-win-win-win for YouTube and other social media applications. Content creators would win. Users would win. YouTube would win. Social media applications would win. You would win. Most importantly, the incentive for creativity would win.

End of Chapter 2.

Chapter 3: The Solution

The Solution

The “FDAIG” Setting Icon Button

The “Freedom to Disable Artificial Intelligence - Generated Content” Setting Icon Button

Chapter 3: The Solution

Solution Page

The "FDAIG" setting icon button or the "Freedom to Disable Artificial Intelligence - Generated content" setting icon button.

Through the configuration settings of YouTube, and other social media applications, the "FDAIG" icon button / the "Freedom to Disable Artificial Intelligence – Generated Content" setting icon button could be accessed while the user is signed in or not while signed in.

Chapter 3: The Solution

The "FDAIG" setting icon button would appear in large bold letters in the upper or lower corner of YouTube or the social media application's screen interface EG. The icon would be able to be accessed on YouTube & other social media applications by consumers while they are signed in or while they are not signed in.

Example of the "FDAIG" Setting Icon In Action

YouTube **FDAIG** Or Other Social Media Applications Screen Interface

Apon clicking the "FDAIG" setting icon button / "the Freedom to Disable Artificial Intelligence – Generated Content" icon button, users could customize their settings and decide what kind of synthetic content they disable.

Chapter 3: The Solution

Upon clicking the "FDAIG"/ the "Freedom to Disable Artificial Intelligence – Generated Content" setting icon button on YouTube or other social media applications, consumers could either disable all fake content or disable specific synthetic settings:

Disable All Synthetic Artificially Generated Content	Click to Disable
Disable Synthetic Artificially Generated Scripts	Click to Disable
Disable Synthetic Artificially Generated Videography	Click to Disable
Disable Synthetic Artificially Generated Photography	Click to Disable
Disable Synthetic Artificially Generated Voices	Click to Disable

All of these configuration settings within the "FDAIG" setting icon button, can be fully customized. You could click to disable all artificial intelligence-generated content, or you could just disable synthetic scripts, just synthetic voices, just synthetic photography, or just disable synthetic videography.

Draft of how the "FDAIG" Setting Icon Button would appear on YouTube's user Interface (see the arrow)

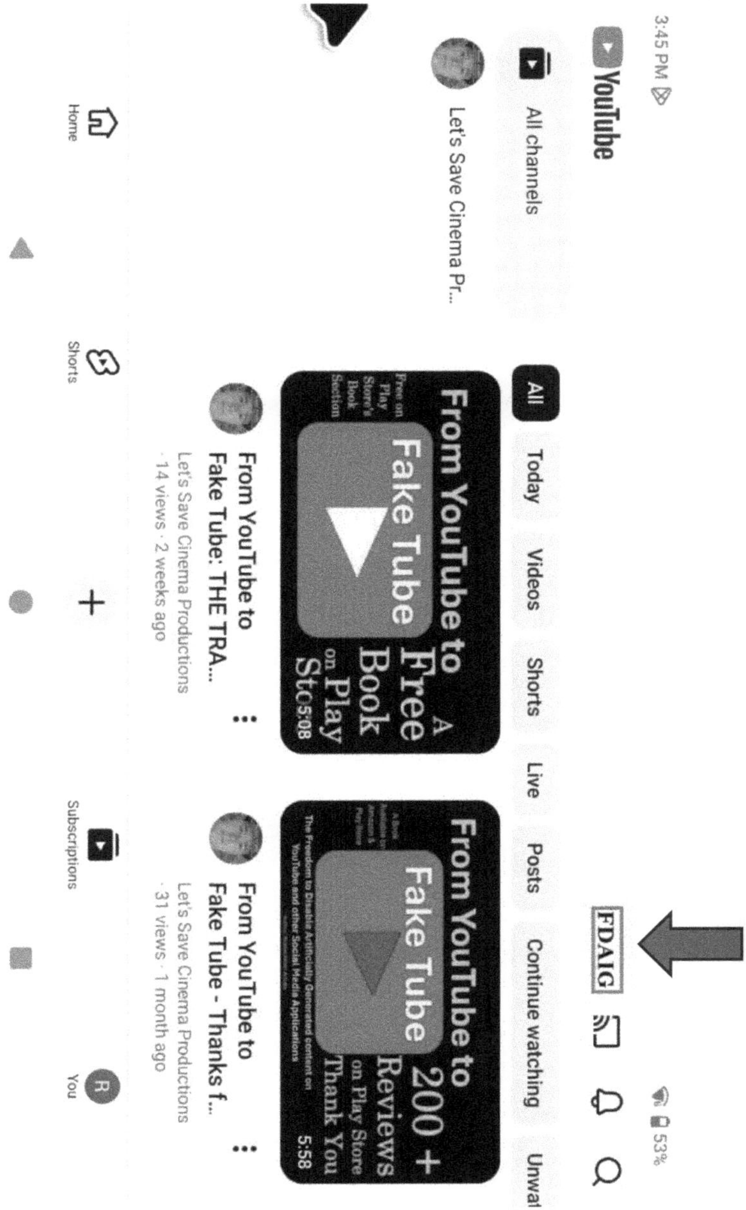

Draft of how the "FDAIG" Setting Icon Button would appear on YouTube's user Interface (see the arrow)

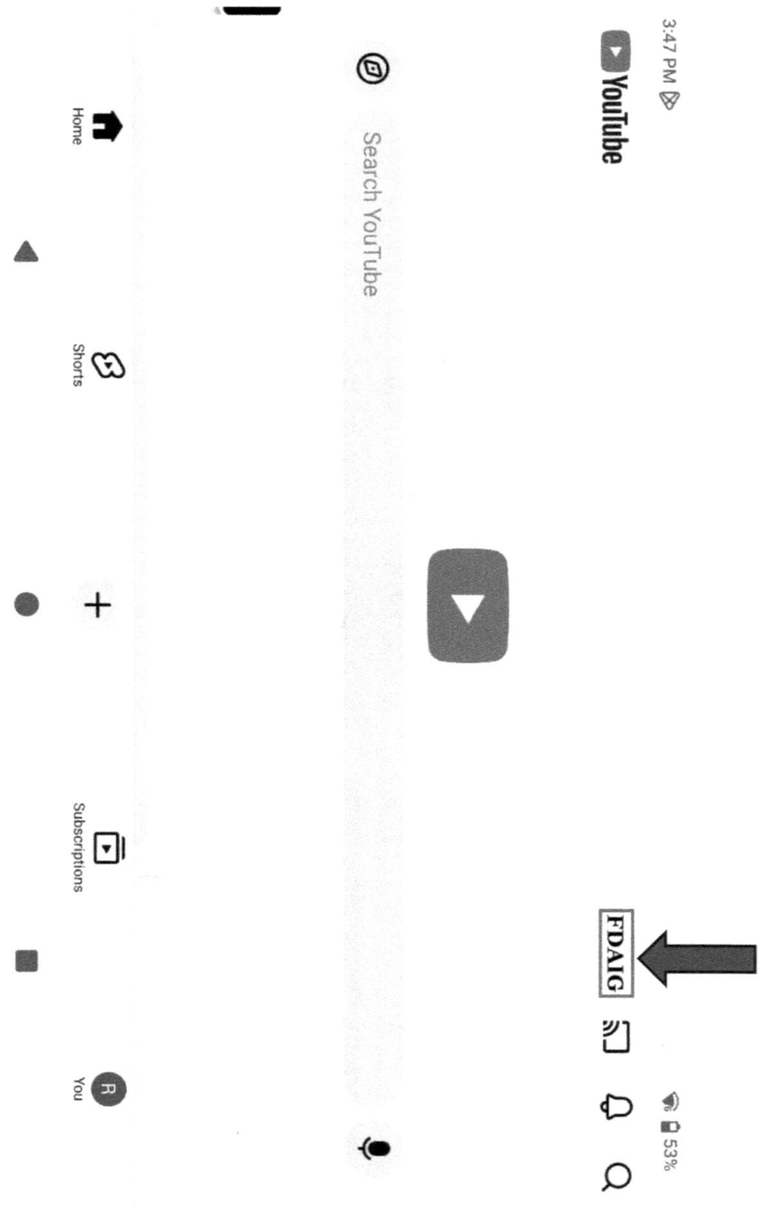

Draft of how the "FDAIG" Setting Icon Button would appear on YouTube's user Interface (see the arrow)

Draft of how the "FDAIG" Setting Icon Button would appear on YouTube's user Interface

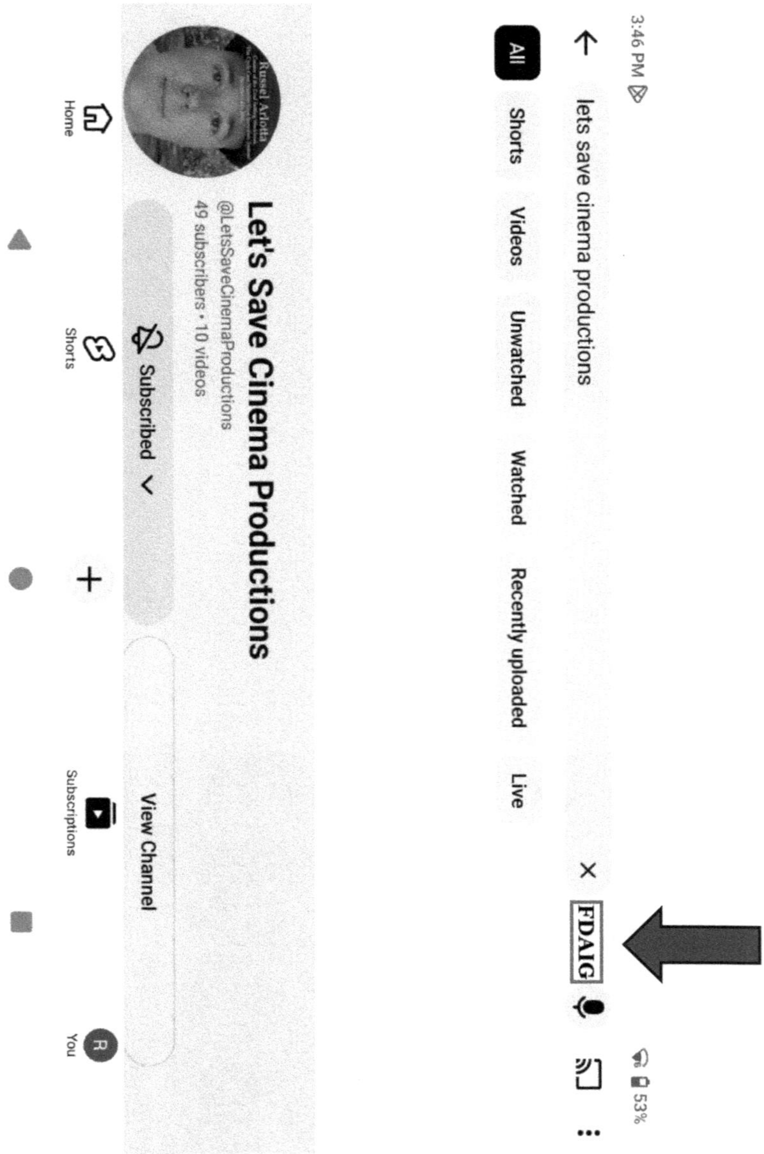

Solution Summary

Upon clicking the "FDAIG" setting icon button on the YouTube user interface, the customizable settings of disabling artificially generated content would appear:

Disable All Synthetic Artificially Generated Content	Click to Disable
Disable Synthetic Artificially Generated Scripts	Click to Disable
Disable Synthetic Artificially Generated Videography	Click to Disable
Disable Synthetic Artificially Generated Photography	Click to Disable
Disable Synthetic Artificially Generated Voices	Click to Disable

The "FDAIG" setting would be accessible whether you are signed in or not signed in to an account.

In Example: Once the "FDAIG" setting icon button is clicked, let's say the "Disable All Synthetic Artificially Generated Content" setting option, all forms of artificially generated videos would NOT appear in your Homepage's algorhythm, and all forms of artificially generated videos would NOT appear in your related videos' algorhythm. This setting would allow actual human content creators' videos to be more easily found and appreciated by others.

Note: Once the "FDAIG" icon setting button is clicked and configured to "Disable All Synthetic Artificially Generated Content" on other social media applications, all forms of artificially generated content would not appear in your timeline (I.E Facebook, X, Instagram).

Solution Summary

Alternative "FDAIG" Setting Options: Disable Specific Types of Synthetic Artificially Generated Content

In Example: Once the "FDAIG" setting icon button is clicked, let's say the "Disable Synthetic Artificially Generated Script" setting option, forms of artificially generated videos that are made using synthetic artificially generated scripts, would NOT appear in your Homepage's algorhythm, and all forms of videos made with artificially generated scripts would NOT appear in your related videos' algorhythm.

Disable All Synthetic Artificially Generated Content	Click to Disable
Disable Synthetic Artificially Generated Scripts	Click to Disable
Disable Synthetic Artificially Generated Videography	Click to Disable
Disable Synthetic Artificially Generated Photography	Click to Disable
Disable Synthetic Artificially Generated Voices	Click to Disable

Users could either select the first option of disabling all synthetic artificially generated content from appearing on their home pages & related videos' algo-rhythms or they could just select to disable certa types of artificially generated videos from appearing on their home pages & related videos' algo-rhythms (I.E Disabling Synthetic Scripts, Synthetic Videography, or Synthetic Photography, or Synthetic Voice

Note: While using the "FDAIG" icon setting button on other social media applications, users could also just disable certain types of artificially genera content from appearing on their' timelines (I.E Facebook, X, Instagram).

Chapter 4: Conclusion

Considering how there are virtually thousands of ways to configure settings on YouTube and other social media applications, the ability to add the "FDAIG" setting icon button is only one more setting option to add. To not provide this simple settings option to users is in, the mildest manners put, not logical, not moral, not practical, and not beneficial to creators & consumers.

The "FDAIG" / the "Freedom to Disable Artificial Intelligence – Generated Content" setting icon button would empower genuine human content creators with an incentive to create high quality real videos. The "FDAIG" icon button would improve the user experience of YouTube and other social media applications, for both content creators and consumers/watchers.

The "FDAIG" setting icon button would also improve the mental health and wellbeing of humans. The "FDAIG" icon button would also be a vital tool to equip parents to shield their children from fake content, and to let their kids just be kids that interact with humanmade content.

Chapter 4: Conclusion

The spamming of fake artificial content on YouTube and other social media applications is inhumane in many senses. Commonsense and freedom of media consumption through optional settings will solve most of the issues addressed in this book.

The solution to provide consumers of YouTube and other social media applications with an experience that allows the freedom to choose to disable artificially generated synthetic videos, is not censorship but rather the solution is achieved with freedom of settings.

When there are innumerable settings offered, but the most important setting of disabling fake content is not offered, that is a blatant violation of freedom of media consumption. Freedom of settings configuration is a part of freedom of media consumption. When the majority of content recommended on YouTube's algorithm is generated by programs, the effect is brutal to the human soul and the human conscience of children, young adults, and adults.

Chapter 4: Conclusion

There has never been a worse time to be alive as a child or young adult (in regard to the mental health of these age demographics) as a result of the invasive and spamming pollution of artificially generated synthetic content on YouTube and other social media platforms.

While artificially generated fake videos spam and ruin the YouTube experience for content creators and consumers, it is also very important to highlight the brutal reality that all this has on young people's minds and wellbeing.

Many independent studies (not done by biased and agenda-ed tech companies) are needed to fully document the amount of negative effects that has occurred to the youth and adults as a result of their brains interacting with artificially generated content on YouTube and other social media platforms.

The majority of suffering cannot be seen, as the human conscious of the mind is invisible. I believe that from now to the end of time, studies will be released that artificial intelligence-created content reduces the quality of mental health and as a result, physical health of adults, young adults, teenagers, and children.

Chapter 4: Conclusion

To ignore such suffering and pretend that this mental suffering (caused by synthetic content) does not exist and that everything is fine - is a disgrace to the human condition and the reality of the human being.

If you think that the minds and hearts of this generation and the next are not negatively affected by artificial intelligence-generated content, I ask you one question:

Would you want your childhood to have been spent being spammed, trashed, and flooded with artificially generated fake videos and content created by unknown parts from unknown locations?

The vast majority of these answers are obviously no.

If people still disagree with the idea of providing an optional icon setting button to disable artificially generated fake videos and content on YouTube and other social media platforms through an optional configuration setting, I ask them to put aside their beliefs and opinions and think of these words:

“Settings are a part of Freedom of Media Consumption”.

Chapter 4: Conclusion

Freedom of settings configurations on YouTube and other social media applications is a vital part of freedom of media consumption.

When you can literally filter out videos and apply hundreds of combinations of settings when you use YouTube or other social media platforms, yet you can't apply a setting that disables artificially generated videos, you simply do not have access to freedom of media consumption. Freedom of consumption includes freedom of configuration and freedom of settings.

Putting opinions aside, putting beliefs aside, the "FDAIG" / the "Freedom to Disable Artificial Intelligence – Generated Content" setting icon button is fundamentally an optional freedom of media settings - choice.

As humans, we deserve the right to configure our settings on YouTube and other social media applications (while being signed in or not signed in), to disable synthetic videos and content, whether it is synthetic scripts, synthetic photos, or synthetic videos. This is not a political debate or a religious debate, this is a debate of freedom of settings and common sense.

Chapter 4: Conclusion

The common arguments against this opinion are non-sensical, anti-freedom of media consumption, and primarily consist of personal generalizations and attacks. The common responses to the topic of this book, is general attacks by labeling those who agree with opinions expressed in this book as being of a certain political, religious, or ideological stock.

The "FDAIG" setting icon button idea is perfectly in line with freedom of media consumption and is supported by a wide range of people regardless of religious backgrounds, regardless of political backgrounds, and regardless of social backgrounds.

The "Freedom to Disable Artificial Intelligence -Generated Content" setting icon button is a universally (and I believe majority) agreed upon freedom of media consumption - setting option regardless of religious, political, or cultural backgrounds. Configuration of settings is a part of freedom of media consumption. I believe that all humans deserve freedom of media consumption and the optional choice to disable fake videos on YouTube and other social media applications.

Chapter 4: Conclusion

People deserve to choose what kind of content they want to consume, and to believe otherwise is anti-consumer, anti-choice, and anti-freedom.

I hope that one day the "FDAIG" icon button, or the "Freedom to Disable Artificial Intelligence – Generated Content" setting icon button can be required by the FCC and that all major internet government bodies will require applications to display the "FDAIG" icon button on the upper or lower corners of YouTube and other social media applications.

To conclude my conclusion, the proposed "FDAIG" setting icon button would be a win-win-win-win.

YouTube would WIN.

YouTube content creators would WIN.

YouTube consumers/watchers would WIN.

You would WIN and...

instead of being 'Fake Tube'/'Robot Tube'

YouTube would return to being YouTube

Dedications: I thank you for reading this book,

I dedicate this book to the youth and future youth, in the hopes that they can enjoy quality content created by humans for humans (and not content generated by programs that simply spam algorithms and destroy the user experience, cause headaches, and cause mental stress).

Life is a gift, and humans, especially young humans, deserve the right to develop and interact with YouTube and other social media applications without fake artificially generated content spamming and littering their digital experiences, if they so choose.

I was privileged to enjoy the golden age of YouTube beginning in 2006 which lasted for over a decade. It is with the creation of this book, that I hope the youth and the future generations of youth to experience meaningful, purposeful, and uplifting experiences through YouTube and other social media applications.

This book is available for free on digital platforms and will be free on additional platforms later.

If you enjoyed this book, please feel free to tell your friends and family about this book.

Enjoy the wealth of the literature of the past (and present) to experience and remember what it means to be human. Never let peoples' attacks on your character waiver your opinions on what you think is moral and right.

Russel Alden Arlotta -Creator of the 2013 Goal setting workbook, 'The Cycle Case Number Goal Invention System'

-Creator of the 2020 Textbook, *Goal Invention: The Mental Therapy, Sport, and Hobby*

Author Bibliography Complete Bibliography

1. Goal Invention: The Mental Therapy, Sport, and Hobby 2021
2. How Cellphones Cropped, Distorted, and Visually Censored Cinema: Let's Save Cinema 2021
3. The Detective Show Directory 2022
4. Light Sensitivity Pain Reduction Techniques for Electronic Devices 2023
5. The U.K. Comedy Show Directory 2023

About the Author

Russel Alden Arlotta founded and created a goal setting workbook in 2013 in a drawing journal, which he called "The Cycle Case Number Goal Invention System"; However, he only used the goal setting workbook for personal use from 2013 – 2019, as he developed his workbook system in drawing journals. The development of his goal setting system is what moved him to choose his pseudonym to be named after it, I.E, "The Cycle Case Number Goal Invention System Author", a pseudonym he uses on all his written projects.

In 2019, he started his writing career, when he began turning his goal setting workbook system, "The Cycle Case Number Goal Invention System", into a full-length reader interactive book, called, "Goal Invention: The Mental Therapy, Sport, and Hobby". His book "Goal Invention: The Mental Therapy, Sport, and Hobby", was finished at the end of 2020 and released in Quarter 1 of 2021 through Politics & Prose's 'Opus Self-Publishing' publishing company.

Russel Arlotta's second book, called, "How Cellphones Cropped, Distorted, and Visually Censored Cinema: Let's Save Cinema", is a book that he began working on during 2020, and he released the first edition of the book in Quarter 2 of 2021 through Politics & Prose's 'Opus Self-Publishing' platform. Russel Arlotta continued to work on his second book full-time throughout the remainder of 2021 and released the second edition of "How Cellphones Cropped, Distorted, and Visually Censored Cinema: Let's Save Cinema" in Quarter 4 of 2021 through Politics & Prose 'Opus Self-Publishing.

It is important to note that Russel Arlotta, later released the third edition of "How Cellphones Cropped, Distorted, and Visually Censored Cinema: Let's Save Cinema" in 2024. The 3rd edition of that book released, Section 6, "The Movie Streaming Company Friendly

Wording Version of the Cinema 3-Step Solution". In Section 6, he teaches all of his prior ideas in a wording that does not use critical or negative wordings, thus allowing his ideas to be more friendly and accessible to consumers and movie streaming companies.

In 2022, Russel Arlotta released his third publication, a short book called, "The Detective Show Directory". This was Russel Arlotta's debut to the short book format. This book allows readers to keep track and interact with the detective shows featured in the directory.

In 2023, Russel Arlotta released his fourth book, again in a smaller book format, called, "Light Sensitivity Pain Reduction Techniques for Electronic Devices". His book underwent many minor edits in 2023 and 2024. The book, "Light Sensitivity Pain Reduction Techniques for Electronic Devices", chronicles all the techniques that he used to reduce eye pain while using electronic devices. His book teaches many methods he uses to reduce eye pain while using electronic devices.

Russel Arlotta also released a short book in 2023, called, "The U.K. Comedy Show Directory". He wrote this book to chronicle all of his favorite comedy shows from the United Kingdom. This short book was inspired from his 2022 directory, "The Detective Show Directory". This directory allows readers to monitor and interact with comedy shows created from the United Kingdom. "The U.K. Comedy Show Directory", underwent minor edits in 2023 and 2024.

In 2025, Russel Arlotta launched his promotional campaign to promote his books through a YouTube channel called "Let's Save Cinema Productions". In the summer of 2025 he wrote this book.

www.ingramcontent.com/pod-product-compliance
Lightning Source LLC
LaVergne TN
LVHW050939210926
848482LV00013B/680
* 9 7 9 8 9 9 0 3 0 1 2 3 8 *